Chocolate Chip Cookies

THE STORIES AND THE SCIENCE

WRITTEN BY **Lynnae W. Allred**

© 2025 Lynnae W. Allred

All rights reserved.

No part of this book may be reproduced in any form whatsoever, whether by graphic, visual, electronic, film, microfilm, tape recording, or any other means, without prior written permission of the publisher, except in the case of brief passages embodied in critical reviews and articles.

Hardback ISBN-13: 978-1-7379746-7-3
Paperback ISBN-13: 978-1-7379746-3-5
eBook ISBN-13: 978-1-7379746-5-9

Library of Congress Control Number: 2024914529

Photographs by Shandee Lynn Lott
Cover and interior layout and design by Shawnda Craig
Edited by Becky Ross Michael

Printed in the United States of America
10 9 8 7 6 5 4 3 2 1
Printed on acid-free paper

Image Credits
Public Domain photographs sourced from Wikimedia Commons:

"Auguste Escoffier" (Public domain in the United States. Source: https://commons.wikimedia.org/wiki/File:Auguste_Escoffier_01.jpg)

"Leonard Thompson, First Insulin Patient" (Public domain in the United States. Source: https://commons.wikimedia.org/wiki/File:Leonard_Thomson_patient.png)

"Caroline S. Brooks" (Public domain in the United States. Source: https://commons.wikimedia.org/wiki/File:Caroline_S.Brooks-DPLA-8f20cb3cb6ff2734ef28873e5d5ba8b7(page_1).jpg)

"The Dreaming Iolanthe, King René's Daughter, by Caroline S. Brooks" (Public domain in the United States. Source: https://commons.wikimedia.org/wiki/File:The_Dreaming_Iolanthe,_King_Rene%27s_daughter,_by_Henrich_Herz._A_study_in_butter_by_Caroline_S._Brooks,_from_Robert_N._Dennis_collection_of_stereoscopic_views_2.jpg)

"Edmond Albius" (Public domain in the United States. Source: https://commons.wikimedia.org/wiki/File:Edmond-Albius-Antoine-Roussin-modifi%C3%A9.png)

"Joseph Goldberger" (Public domain in the United States. Source: https://commons.wikimedia.org/wiki/File:Goldberger,_Joseph.jpg)

"Mahatma Gandhi, Studio, 1931" (Public domain in the United States. Source: https://commons.wikimedia.org/wiki/File:Mahatma-Gandhi,_studio,_1931.jpg)

"If we have ever baked chocolate chip cookies together, this book is for you. Stay curious!"
—Grandma Nae

"A toast to NYC, birthdays, and forced goals. Love You!"
—Shandee

"Cheers to those who love eating cookie dough and cookies as much as I do!"
—Shawnda

"NO ONE IS BORN A GREAT COOK. ONE LEARNS BY DOING."

—JULIA CHILD

COOKIE SCIENCE

⟫⟫ THE SCIENCE OF ⟪⟪
CHOCOLATE CHIP COOKIES

People who can measure ingredients and follow recipe instructions carefully will become good cooks. But people who understand the science of food can experiment and combine ingredients in new, interesting, delicious ways. They are called chefs. A chef can create something magical with just a few ingredients and some creativity.

> **A COOK FOLLOWS A RECIPE.**
> **A CHEF CREATES A RECIPE.**
> Scientists who think like chefs change the world.

First, we will learn the science behind how to bake cookies. Next, if you want to know more, you can read the stories of everyday creators and scientists who used their curiosity to "think like chefs." They helped develop the ingredients and processes that made the invention of chocolate chip cookies possible.

AUTHOR'S NOTE:

Baking cookies together is a time-honored tradition that results in more than just a sweet treat–you get sweet memories and relationships too.

Enjoy some of the stories and the science behind chocolate chip cookies, and as you take a closer look, you'll be helping your child lean into their innate curiosity.

THE INVENTOR CHEF
RUTH WAKEFIELD

Ruth Wakefield is widely credited with "inventing" chocolate chip cookies.

Ruth and her husband owned a restaurant known as the Toll House Inn, and her customers loved the delicious food and desserts she made. In 1935, she decided to experiment with one of her customers' favorite recipes, a butterscotch pecan cookie. She wanted to try adding baking chocolate to the batter, but she didn't have any on hand. Instead, she used an ice pick to break up a semi-sweet chocolate bar into little bits. She added these small pieces of chocolate to the cookies, and people loved them. Her customers thought the cookies were so delicious that they started asking for copies of the recipe.

So, Ruth added the recipe to a cookbook she sold to customers who ate in her restaurant, the Toll House Inn. She called the recipe "Toll House Chocolate Crunch Cookies." People started buying more and more chocolate bars from their local grocery stores so they could try Ruth's delicious recipe at home.

When the chocolate manufacturer noticed the surge in chocolate sales in and around Whitman, Massachusetts, they sent someone to investigate. Ruth explained and asked the company if they could make their chocolate bars with scored lines so people could more easily cut chunks of chocolate for their cookie batter.

After a while, the chocolate manufacturer began making smaller drops of chocolate they called "chocolate morsels." You know them as chocolate chips. Ruth gave the company permission to print the recipe and market their semi-sweet chocolate as a key ingredient. In exchange, she received a $1 payment for recipe rights and a lifetime supply of baking chocolate!

That is the story of how one chef invented a whole new recipe that is loved all over the world. Are you ready to become a chef and a cookie scientist like Ruth?

FIRST, WASH YOUR HANDS,
THEN GATHER YOUR INGREDIENTS AND TOOLS.

You wouldn't bother putting together a puzzle if you knew a few of the pieces were missing, would you? The whole puzzle would be ruined.

What if you started cooking something and then realized you didn't have all the ingredients? Professional chefs make sure they have mise en place (pronounced "meez ahn plahs") before they begin. Mise en place is a French phrase that means "everything in its place." Gathering all your ingredients and tools before you start to cook assures you have everything you need.

You'll be a more efficient culinary scientist (someone who studies food and cooking) if you develop the habit of mise en place every time you prepare a dish or meal. It's the first step in learning how to think like a chef.

Note: You can measure out each of your ingredients first and set them in small individual bowls so all of your ingredients are measured and ready before you start mixing. **LEARN ALL ABOUT THE CHEF WHO INNOVATED THE PRACTICE OF "MISE EN PLACE" ON PAGE 31.**

INGREDIENTS:
- Granulated Sugar
- Brown Sugar
- Butter
- Eggs
- Vanilla
- Baking Soda
- Salt
- Flour
- Chocolate Chips

TOOLS:
- Mixing bowl
- Measuring cups
- Measuring spoons
- Electric mixer and beaters or cookie paddles (hand mixer or stand mixer)
- Silicone spatula
- Parchment paper, silicone mat, or oil spray
- Cookie sheet
- Kitchen scale (optional)

COOKIE SCIENCE

>>> ONE CUP <<<

All cookie scientists need to be good at measuring. A set of measuring cups comes with different-sized cups. Try an experiment: If you fill a ¼ cup full of sugar, how many of those will it take to fill the 1-cup size clear to the top?

TRY IT!

One cup will hold exactly four ¼ cup measures or three ⅓ cup measures or eight ⅛ cup measures.
IT'S MATH!

LET'S BEGIN WITH THE FIRST INGREDIENT, GRANULATED SUGAR...

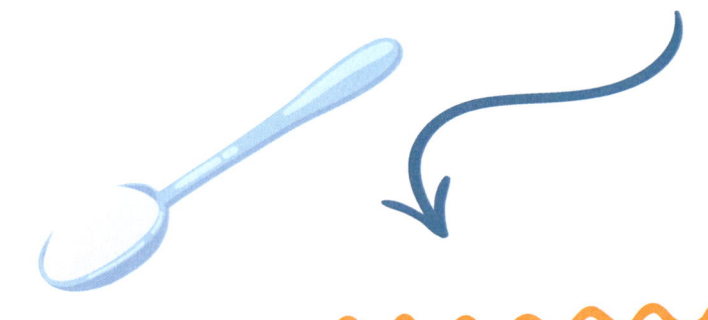

1- MEASURE 1 CUP OF GRANULATED SUGAR
2- ADD IT TO YOUR MIXING BOWL

1

2

Q: WHY DON'T VAMPIRES LIKE SUGAR?
A: THEY PREFER NECK-TARINES!

COOKIE SCIENCE

Most English-speaking countries use one of two measuring systems: the customary system (also known as the imperial system) or the metric system.

The U.S. typically uses the customary system in recipes even though the metric system can be more precise. Why would Americans do that? Well, the answer might be…PIRATES.

It's a great story and you can read all about it on page 32 of this book. Meanwhile, if you are a cookie scientist, you can try baking cookies using metric units and see which system you like best.

CUSTOMARY RECIPE:
- 1 cup granulated sugar
- 1 cup brown sugar
- 1 ¼ cups soft butter
- 2 eggs
- 1 teaspoon vanilla extract
- 3 cups all-purpose flour
- 1 teaspoon salt
- 1 teaspoon baking soda
- 1 ½ cups chocolate chips

METRIC RECIPE:
- 220 grams granulated sugar
- 200 grams brown sugar
- 280 grams soft butter
- 2 eggs
- 1 teaspoon vanilla extract
- 375 grams all-purpose flour
- 1 teaspoon salt
- 1 teaspoon baking soda
- 270 grams chocolate chips

THERE'S MORE TO LEARN ABOUT SUGAR ON PAGES 33–35.

IF YOU HAVE A KITCHEN SCALE, PRACTICE MEASURING SOME OR ALL OF YOUR INGREDIENTS BY WEIGHT.

IT'S FUN!

THE SECOND INGREDIENT IS
BROWN SUGAR...

3- MEASURE 1 CUP OF BROWN SUGAR
4- PACK IT DOWN
5- ADD THAT TO YOUR MIXING BOWL TOO

8 | Chocolate Chip Cookies

COOKIE SCIENCE

THERE ARE MANY KINDS OF SUGAR, including granulated and brown.

Brown sugar contains molasses, which gives it a darker color and a richer flavor. It contains more moisture content too. We include brown sugar in this recipe to help keep our cookies soft and moist even after they are finished baking.

"FEEL GOOD" NEUROTRANSMITTERS

WHEN WE EAT COOKIES WITH SUGAR, our brain releases a chemical called dopamine.

DOPAMINE is a **"FEEL GOOD" NEUROTRANSMITTER** that helps us feel happy and satisfied. Our brains like to repeat that experience, and we want to eat even more cookies. Some people call chocolate chip cookies a **"COMFORT FOOD"** because they feel pleasure when they smell or eat them.

If we eat too many sugary sweets, or eat them too often, our brain gets **"STUCK."** We keep looking for a cookie or something sugary to help us feel better when we feel sad, bored, lonely, worried, or tired. It is good for our brains to learn that there are many, many ways to release dopamine so we can start to feel happy and motivated again. Exercising, going out into the sunshine, listening to music, or getting some sleep are just some of the ways to help your brain get more dopamine.

THE THIRD INGREDIENT IS
BUTTER...

6- MEASURE 1 ¼ CUPS OF SOFT BUTTER
*MAKE SURE THE BUTTER IS SOFT BUT NOT MELTED

7- ADD THE BUTTER TO THE MIXING BOWL

8- MIX UNTIL EVERYTHING IS SMOOTH AND CREAMY

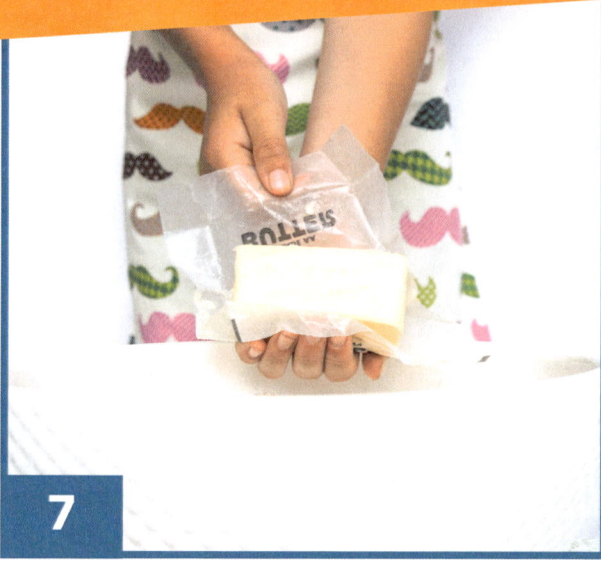

COOKIE SCIENCE

CREAMING BUTTER AND SUGAR

Cookie scientists know a couple of secrets when it comes to butter. First, mixing the butter and sugar together before you add other ingredients to the mixing bowl is called "creaming." Sugar crystals have jagged edges, and these edges drag some air into the soft butter to make little pockets that will fill with carbon dioxide gas as the cookies bake. This will help the cookies rise higher, making them tender instead of heavy and dense.

Second, the temperature of the butter you start with really changes the way your finished cookie looks. Below, you can see cookie samples where the only difference is the temperature of the butter during creaming.

TEMPERATURES OF THE BUTTER LEFT TO RIGHT:

Melted butter (128°F, 53°C)

Room-temperature softened butter (74°F, 23°C)

Cold butter (46°F, 7°C)

Often, when you start with **MELTED BUTTER**, cookies will spread out too flat, almost like they melted. This recipe is quite forgiving, so melted butter didn't ruin these cookies too much. **ROOM-TEMPERATURE BUTTER** made the prettiest cookies. **VERY COLD BUTTER** stays in chunks instead of creaming into the dough as it should, and when you bake it, it melts out in puddles like lava from a volcano.

READ THE FUN STORIES ABOUT BUTTER ON PAGES 36–37.

THE FOURTH INGREDIENT IS
EGGS...

9- ADD 2 EGGS TO THE MIXING BOWL

Q: WHAT HAPPENS WHEN YOU TELL AN EGG A JOKE?

A: IT CRACKS UP!

COOKIE SCIENCE

EGGS provide protein, fat, and water to your cookie dough.

The protein helps provide structure to your baked cookie, so all of the pieces hold together. Without eggs, your cookies would be flat, crumbly, and gritty.

MORE FUN FACTS ABOUT EGGS ON PAGES 38–39.

– ONE TEASPOON –

JUST LIKE MEASURING CUPS, your different sizes of measuring spoons will help you learn more important math.

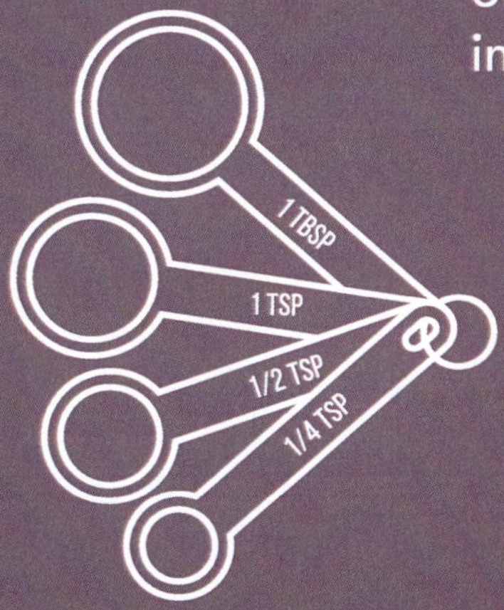

- How many ¼ teaspoons does it take to fill up one teaspoon?
- Trick question: How many teaspoons to fill up one tablespoon?
- How many tablespoons to fill up exactly ¼ cup? Can you experiment to find out?

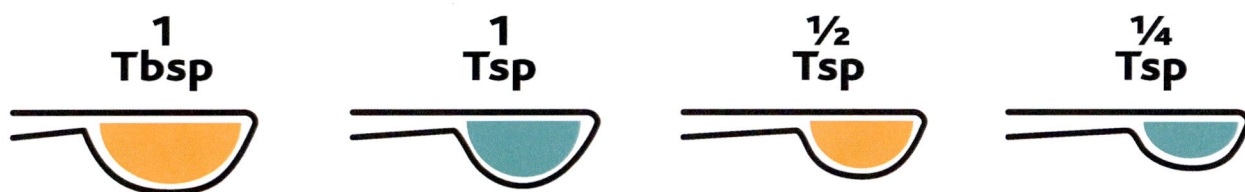

THE NEXT AND FIFTH INGREDIENT IS
VANILLA...

10- MEASURE 1 TEASPOON OF VANILLA

11- ADD THE VANILLA TO THE MIXING BOWL

12- MIX EVERYTHING AGAIN

COOKIE SCIENCE

If you are adding **REAL VANILLA** to your cookie dough, you just might be rich! Well, rich compared to many people in the world. This is because real vanilla is the second most expensive spice in the world, with saffron being the most expensive.

What you may actually be using in your kitchen is known as "imitation vanilla." It still smells and tastes good, but it doesn't have all of the flavor compounds found in real vanilla because it is created through a chemical process and not grown on a vine.

REAL VANILLA VERSUS IMITATION VANILLA

WHY IS REAL VANILLA SO EXPENSIVE AND SO RARE?
STORY COMING UP ON PAGES 40–41.

THE SIXTH INGREDIENT IS
FLOUR...

13- MEASURE 3 CUPS OF ALL-PURPOSE FLOUR
14- ADD THE FLOUR TO THE MIXING BOWL

Q: HOW DO YOU MAKE FRENCH BREAD?

A: WITH EIFFEL FLOUR.

COOKIE SCIENCE

FLOUR IS MADE BY grinding a hard kernel of wheat into a powder. Wheat berries look like fat brown grains of rice. Humans have been grinding wheat and other grains into flour for centuries.

It's only been for about the last 150 years that we've been doing something to flour that is a little weird. We have been making it less nourishing. On purpose!

THERE'S A STORY ABOUT THAT ON PAGES 42–45.

THE SEVENTH INGREDIENT IS SALT...

15- MEASURE 1 TEASPOON OF SALT
16- ADD THE SALT TO THE MIXING BOWL

15

16

Q: WHAT'S A PIRATE'S FAVORITE SEASONING?

A: SEA SALT-ARRRR!

18 | Chocolate Chip Cookies

COOKIE SCIENCE

THAT SMALL TEASPOON OF SALT plays a critical role in your cookie dough. Salt helps intensify all of the flavors in your dough. Without it, your cookies will be bland and flavorless. They will become dry and stale faster because salt also acts as a preservative.

SUPER POWERS OF SALT!

SALT IS MADE OF a combination of two elements, **SODIUM AND CHLORIDE.**

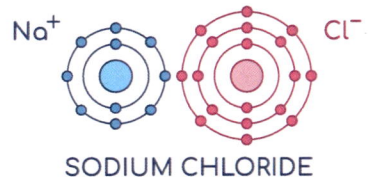

SODIUM CHLORIDE

Every cell in your body requires these elements. Without sodium, you could not contract and relax your muscles because your nerves wouldn't be able to pass impulses from your brain to other parts of your body. You couldn't maintain the proper amount of fluids and minerals in your body without sodium. Your body wouldn't be able to heal when it is injured.

You couldn't survive more than a week or two if your body didn't have a way to get salt regularly. And if you think only your cookies depend on salt, you have really underestimated salt. Entire wars have been fought over which governments and rulers had control over the salt supply.

BUT THAT'S ANOTHER STORY... SEE PAGES 46–47 FOR MORE SALTY INFO!

THE EIGHTH INGREDIENT IS
BAKING SODA...

- 17- MEASURE 1 TEASPOON OF BAKING SODA
- 18- ADD THE BAKING SODA TO THE MIXING BOWL
- 19- MIX EVERYTHING AGAIN

17

18

19

COOKIE SCIENCE

BAKING SODA IS A "LEAVENING" AGENT IN COOKIES.
That means it mixes with acids from the brown sugar, eggs, and chocolate to start a chemical reaction when the dough is heated. The heat causes the baking soda to break down, giving off little bubbles of carbon dioxide gas. The gas gravitates to the small air pockets you made earlier by creaming the sugars and butter. Water from the eggs forms steam. The bubbles of steam and gas make your cookies rise higher, so they are light and delicious.

LEARN MORE ABOUT BAKING SODA ON PAGE 48.

BAKING SODA MAGIC

You may have created a **KITCHEN VOLCANO** with **BAKING SODA** and **VINEGAR** before. That's a fun experiment that shows how mixing sodium bicarbonate (baking soda) with an acid in vinegar will create carbon dioxide bubbles that expand and cause your volcano to **"ERUPT."**

BUT THAT'S NOT ALL! Baking soda is a little bit gritty when you add water, and that abrasive texture makes it ideal for household cleaning. It can scrub dirt, deodorize stinky shoes, and lift stubborn stains.

BAKING SODA BALLOON EXPERIMENT

Follow the illustration below for the experiment.
You will need: baking soda, vinegar, a balloon, and a funnel.

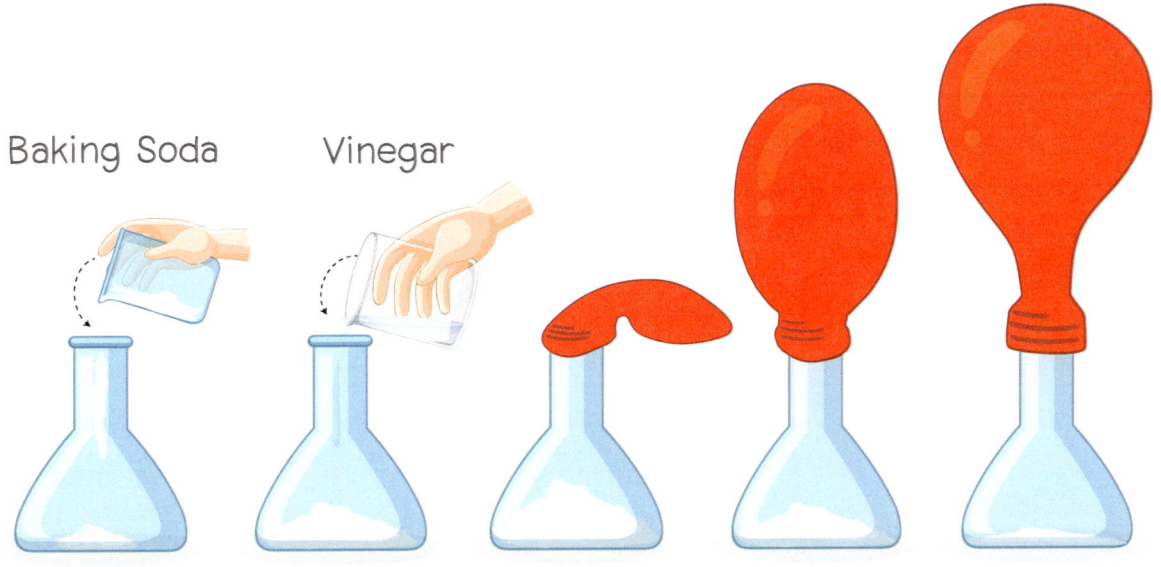

THE NINTH AND LAST INGREDIENT IS
CHOCOLATE CHIPS...

20- MEASURE 1 ½ CUPS OF CHOCOLATE CHIPS

21- ADD THE CHOCOLATE CHIPS TO THE MIXING BOWL

20

21

Q: WHAT FRUIT LOVES CHOCOLATE?

A: A COCOA NUT.

22 | Chocolate Chip Cookies

COOKIE SCIENCE

Remember the **"FEEL GOOD" NEUROTRANSMITTER** in sugar?
Chocolate contains a chemical called tryptophan.

"FEEL GOOD" NEUROTRANSMITTERS

Our brains use **TRYPTOPHAN** to produce another **"FEEL GOOD" NEUROTRANSMITTER CALLED SEROTONIN.**

This brain chemical helps us feel happier and more joyful. Just smelling chocolate can produce brain waves that help us feel more relaxed. Between the sugar and the chocolate in cookies, we get a double dose of brain chemicals that help us feel good. **NO WONDER WE LOVE CHOCOLATE CHIP COOKIES SO MUCH!**

LEARN MORE ABOUT CHOCOLATE ON PAGES 49–50.

The Stories and the Science

ONE LAST MIX.
CHOCOLATE CHIP COOKIE DOUGH!

22- MIX YOUR COOKIE DOUGH ONE LAST TIME

Q: WHAT DO YOU CALL A COOKIE DOUGH THAT DANCES REALLY WELL?

A: ONE SMOOTH COOKIE!

COOKIE SCIENCE

CHILL THE DOUGH

Your cookies will taste best if you **CHILL THE DOUGH** in the refrigerator for a day. This helps the flavors intensify, plus the fat in the butter is kept cold so your cookies stay round and spread out less. But if you want to bake them right away, you may.

GET READY TO
BAKE YOUR COOKIES!

23- HEAT THE OVEN TO 375°FAHRENHEIT (190°CELSIUS)

24- USE A COOKIE SCOOP OR A SPOON AND SCOOP THE COOKIE DOUGH INTO ROUNDED BALLS ONTO THE BAKING SHEET

*USE COOKING SPRAY, PARCHMENT PAPER, OR A SILICONE MAT ON THE BAKING SHEET TO PREVENT YOUR COOKIES FROM STICKING

23

24

TIDY UP: NOW IS A GOOD TIME TO CLEAN UP YOUR MEASURING TOOLS AND INGREDIENTS WHILE THE OVEN HEATS UP.

Chocolate Chip Cookies

COOKIE SCIENCE

BEFORE YOU LICK THE SPOON...

Most people know it isn't a good idea to eat raw eggs. But did you know that raw flour can carry harmful bacteria like salmonella and E. coli too? Flour is made from raw grain, and sometimes the grain becomes contaminated by bacteria while it is in the field or while it is being processed and milled into flour.

All recipes with raw flour and raw eggs must be cooked before we eat them to kill these bacteria. Otherwise, you risk getting sick. It's also important to clean all tools and surfaces with warm, soapy water when you finish baking.

AND NOW...
LET'S BAKE!

25- BAKE YOUR COOKIES FOR 8–10 MINUTES

26- WHEN YOUR COOKIES ARE FINISHED BAKING, USE AN OVEN MITT TO REMOVE THE BAKING SHEET FROM THE OVEN

27- LET THE COOKIES COOL ON THE BAKING SHEET FOR 10 MINUTES, THEN TRANSFER TO A WIRE RACK TO COOL COMPLETELY AND ENJOY!

25

26

27

COOKIE SCIENCE

READ ON TO LEARN MORE!

EACH INGREDIENT HAS A STORY. When you know the stories and the science, you'll be one step closer to becoming a true cookie scientist!

≫≫ THE MAILLARD REACTION

Once you slide your cookies into the oven, some exciting things start to happen. When your cookies reach 280°F (140°C), amino acids from the egg and the sugars in your dough combine to create new flavor molecules. They also cause the cookies to start turning a darker brown color. This is called the Maillard (pronounced my-yard) reaction.

A good chef knows how to use the Maillard reaction to produce the very best flavors in food, including meats and vegetables. If you have ever toasted a bagel, browned potatoes to make french fries, or toasted a marshmallow over an open fire, you are using the Maillard reaction to make your food delicious.

Cookies are finished baking when they start to turn a little brown on the edges. They aren't shiny anymore. Take them out of the oven now, and they will finish baking on the cookie sheet. For the softest cookies, move them to a cooling rack right away.

≫≫ CARAMELIZATION

When your cookie dough reaches a temperature of precisely 356°F (180°C), a second chemical reaction called "caramelization" starts. The chemical compounds in your cookies change and give off a nutty, sweet fragrance.

Caramelization won't happen if your cookies don't get warm enough. That is why we heat the oven to 375°F (190°C). If you are paying close attention, your nose will be almost as accurate as a kitchen timer, telling you when your cookies are ready.

THE STORIES
OF GREAT SCIENTISTS AND CHEFS

MISE EN PLACE

AUGUSTE ESCOFFIER

A French chef, Auguste Escoffier, was the first to teach mise en place, or "everything in its place." He served for a time as an army chef with the responsibility to feed hundreds of soldiers at once. He learned that it was important to be efficient in the way the kitchen was set up.

In those days (the late 1800s), most restaurant kitchens were dirty and disorganized. Escoffier taught the importance of cleanliness and sanitation. In his restaurant, there was a chain of command, just like in the army. Each person had specific duties. Each person also had all of the tools and ingredients necessary to make food quickly and make it precisely the same way every time. Escoffier knew that well-trained chefs would create exceptional food.

Auguste Escoffier revolutionized the restaurant industry with his ideas, but he also cared about people. He found ways to share unused and leftover food with people who didn't have enough to eat. He taught that food tasted best if it could be purchased directly from the farmers who grew it. He worked to set up programs to make it possible for his kitchen workers to have medical care and to retire with a pension.

He wrote down and preserved thousands of recipes, and many are still famous today. Escoffier was a chef, not just a cook. His thoughtful, inventive ideas changed the way we prepare and eat food, and his systems are still practiced all over the world.

DID YOU KNOW...?

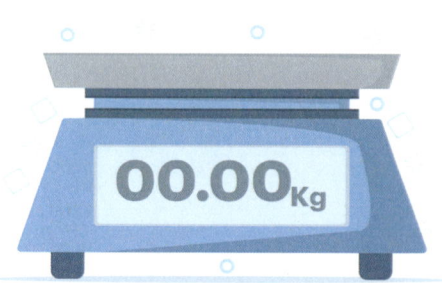

CUSTOMARY MEASURING VERSUS METRIC MEASURING

When **THE UNITED STATES OF AMERICA** was a brand-new country, people wanted to set up a system so everyone in the new Republic would measure things the same way.

In 1793, Secretary of State, **THOMAS JEFFERSON**, wrote to his friends in France who were developing a new system of measurement. These friends sent a scientist named **JOSEPH DOMBEY** to the United States by ship. With him, he carried a funny-looking copper cylinder with a little handle on top. This cylinder was actually a very precise **KILOGRAM WEIGHT,** which was supposed to help **JEFFERSON** set up the **METRIC SYSTEM** in the U.S.

Unfortunately, the ship Dombey was traveling on was blown off course by a big storm. It ended up in the Caribbean, where pirates (British privateers) took **DOMBEY** prisoner and auctioned off his belongings. **DOMBEY** died in captivity, and that **KILOGRAM WEIGHT NEVER REACHED JEFFERSON**. This missed opportunity may be part of the reason Americans just kept using the customary system.

Learning a new measuring language can be hard. Americans continue to measure in a way that is familiar to them, even though most people in the rest of the world have converted to metric measurements.

THE STORY OF

>>> SUGAR <<<

THE STORY OF SUGAR IS COMPLICATED. It is one of the world's oldest known commodities (a raw material or agricultural product that can be bought and sold). The first known sugar producers lived in India six thousand years ago. Their sugar came from a plant called sugar cane. Granulated sugar is called sucrose and is an important source of a simpler sugar called glucose, which gives our bodies energy. People who made sugar cane into sugar crystals could trade sugar for money or other things they needed. Sometimes, the greed for money became more important than treating the workers who helped make the sugar with dignity and kindness. Some workers were even enslaved and worked without being paid.

Today, we have many types of sugar and sugar substitutes, and people still sell these different types of sugar. Keep in mind that people who make money selling sugar have an incentive to make you want to eat plenty of it. That is why you need to think like a chef when it comes to added sugar (sugar added by food manufacturers) in your food.

DID YOU KNOW...?

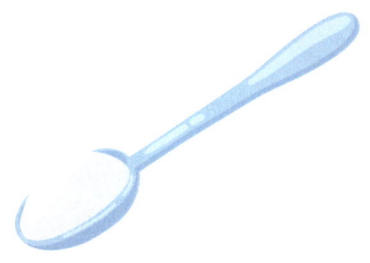

HOW OUR BODIES DIGEST SUGAR

YOU HAVE AN ORGAN IN YOUR BODY CALLED A PANCREAS. One of its main jobs is to produce the hormone, insulin. **INSULIN HELPS OUR BODIES CONVERT THE FOOD WE EAT INTO ENERGY.** If we have too much glucose (sugar) in our blood, our bodies can convert it to fat and store it. Your body can use the stored energy later.

WHAT HAPPENS IF OUR BODIES DO NOT PRODUCE ENOUGH INSULIN?

We develop a disease known as diabetes. Up until the early 1920s, if someone's pancreas suddenly stopped producing insulin (or didn't make enough of it) it was a big problem!

The glucose in the blood would build up to toxic levels. There was no way doctors could help these people. The best way they could think of to manage glucose in the blood was to tell these patients to eat as little food as possible. People with diabetes had to starve to stay alive. They would usually die within a few months or years, no matter how carefully they ate.

It was a dreadful thing to develop this disease!

THE STORIES
OF GREAT SCIENTISTS AND CHEFS

A YOUNG MAN NAMED LEONARD THOMPSON was living in a hospital in Toronto, Canada, where doctors were trying to keep him alive. He was 14 years old and was quite tall but weighed only 65 pounds (29 kilos). Most healthy boys that age weigh between 100 and 125 pounds (48-57 kilos).

Scientists and doctors had worked for many years to find the cause of and the cure for diabetes. They learned it was possible to take insulin from a dog's pancreas and use it to lower glucose in another dog's blood. Once they were successful with this step, they still had to figure out how to make insulin therapy safe for humans.

Dr. Frederick Banting, Charles Best, James Collip, and John James Rickard Macleod were some of the scientists who collaborated to help develop the first insulin therapy for diabetes. On January 11, 1922, Leonard Thompson was the first human to receive an injection of insulin.

It did not work as well as his doctors had hoped, but James Collip, a biochemist, worked day and night, experimenting again and again to find a way to purify the insulin. On January 23, 1922, Leonard received a second injection of this more purified form of insulin. His blood sugar improved almost immediately, and he started feeling stronger. He would still need insulin injections every day, but for the first time, doctors had a way to help people who had diabetes live longer, healthier lives. This was miraculous!

Their important work was just the beginning. Thousands of scientists continue to think like chefs, to collaborate and share ideas, and to develop treatments for this disease. One day, it may be possible to cure diabetes! Until then, you can keep exploring how your body uses sugar. You can use this information to help keep your body, brain, and blood healthy. You'll learn what kinds of foods are best for giving your very unique and special body the energy it needs.

THE STORY OF
>>> BUTTER <<<

BUTTER IS A FOOD THAT HAS BEEN AROUND for a very long time. Archaeologists believe that humans began churning butter more than 6,000 years ago. While most cookie scientists are familiar with butter made from the cream in cow's milk, other cultures make butter from the milk of sheep, goats, yaks, buffalo, and even reindeer.

Because butter is hard in cooler temperatures, a unique part of its history is that it can be sculpted into shapes. For example, in Tibet, monks have been sculpting butter since 1409. These sculptures are used as part of religious rituals and celebrations.

THE STORIES
OF GREAT SCIENTISTS AND CHEFS

In 1867, a woman from Arkansas named Caroline Shaw Brooks wanted to find a way to encourage people to buy more of the butter she and her husband produced on their farm. She started to think like a chef. She was very artistic and liked to shape the butter she made into small sculptures using tools like straw and butter paddles. She thought just looking at the butter and seeing how creamy and beautiful it was might help convince people to buy more of it.

CAROLINE got better and better at butter sculpting, and in 1876, she created a butter sculpture for the Centennial Exhibition in Philadelphia. This was an elaborate national fair showcasing American industrialization that celebrated the 100th anniversary of the signing of the Declaration of Independence. The fair lasted for six months. Millions of people saw her interesting artwork made completely from butter.

Dreaming Iolanthe, butter sculpture, 1876 Centennial Exhibition

This started a new trend in butter sculpting, which became popular at state and county fairs. People began carving butter into all kinds of designs, and dairy farmers liked this because it helped sell more butter. Caroline continued to develop her artistic skills. She eventually studied sculpture in Paris and Florence and became a professional sculptor who made more traditional art with marble and clay. But she never gave up sculpting butter.

THE STORY OF

>>> EGGS <<<

CONSIDER HOW YOUR LIFE WOULD BE DIFFERENT IF YOU COULDN'T GET EGGS. No scrambled eggs, pudding, ice cream, pancakes, cakes, mayonnaise or pasta for you! Eggs are a critical component in many of the foods we eat every day. **YOU WOULD REALLY MISS EGGS.**

This actually happened during the gold rush in San Francisco that began in 1848. So many people flocked to the city at once that local farmers couldn't keep up with the demand for fresh food. When demand is high and supply is low, prices go up. Miners in the field would pay as much as $3 per egg. That doesn't sound like that much today, but back then, that was more than one thousand dollars per dozen!

THE STORIES
OF GREAT SCIENTISTS AND CHEFS

In 1849, two enterprising men, **"DOC" ROBINSON** and **ORRIN DORMAN**, started thinking like chefs. They wondered if they could change one ingredient and get a similar result. They got into a boat and floated out to rocky Farallon Island near San Francisco to gather eggs from the nests of the thousands of seabirds who lived on the island. They filled their boat so full of eggs that it almost sank in the rough seas on the way back. They lost more than half of their eggs in the attempt and nearly lost their lives. They decided they would never do it again. But they sold their remaining eggs for so much money that other men decided to try. The greed over eggs was just like the greed over gold. Egg gathering became dangerous, as different groups fought over who had the right to gather seabird eggs to sell.

Unfortunately, these egg-gatherers didn't consider the cost to the birds. And the seabird population couldn't keep up with demand. The result was that the egg hunters nearly destroyed the seabird population near San Francisco before more farmers moved into the area and started selling hen's eggs for a few cents each. Whether it is the eggs in your chocolate chip cookies or the ones that hold your birthday cake together, knowing how critical it is to have access to eggs can help you develop a new respect for them and for the hens (or other birds) that produce them for you.

THE STORY OF

>>> VANILLA <<<

REAL VANILLA IS SO SPECIAL that if you happen to have some to add to your cookies, you should also know the story of why it is so rare.

VANILLA COMES FROM INSIDE A LONG, THIN BROWN BEAN POD. This bean develops and grows after a beautiful white flower, called a vanilla orchid (Vanilla planifolia) is pollinated by two very special species of bees, the Melipona bee or the Euglema bee. In the early 1800s, vanilla orchids were carried to several countries by traders and explorers, but the traders didn't know they needed to take the bees along. No bees. No vanilla bean pods. No vanilla.

THE STORIES
OF GREAT SCIENTISTS AND CHEFS

IN THE EARLY 1840s, ON THE SMALL ISLAND OF RÉUNION (near Madagascar), a 12-year-old enslaved worker named **EDMOND ALBIUS** knew how to think like a chef. He used his experience watching how bees pollinate flowers and substituted a new ingredient into the pollination process: his own hands. He used a small wooden needle or a blade of grass to open a membrane inside the flower. Then, **EDMOND** used his fingers to press two parts of the flower together so they were touching. This is called "hand pollination." His discovery changed everything for the vanilla growers.

Today, much of the world's vanilla is produced in Madagascar, and almost all vanilla orchids must be pollinated by hand by an experienced grower. To complicate things even more, each vanilla orchid blooms only one day out of the year. Growers must hand-pollinate the orchid on the day it blooms, or no vanilla bean will grow. If the flower is pollinated, it takes nine months before a mature vanilla bean can be harvested. Then, it takes another three months for growers to complete a labor-intensive process of curing the bean pods. This difficult process of hand-pollinating, growing, harvesting, and curing vanilla beans makes it the most labor-intensive agricultural crop on the planet. That's why people created a shortcut and started making imitation vanilla.

Before you add your vanilla flavoring to your cookie dough, take just a minute to breathe in its fragrance. Because of Edmond Albius's discovery, and thanks to hard working growers, vanilla is accessible even if you don't live in a tropical climate. It gives your cookies a delicious flavor.

THE STORY OF

»» FLOUR ««

HERE'S WHAT HAPPENS WHEN WE GRIND WHEAT INTO FLOUR:
A grain of wheat has three parts. The bran, the endosperm, and the germ. When the grain is smashed by grinding, the endosperm is white and powdery. The bran and the germ are gritty and brown.

Bakers soon figured out that if they could sift out the hard, brown bran and germ, their baked goods turned out fluffier and were lighter in color. It took extra time to sift out the brown parts of the flour, so bakers had to charge more for their fluffy, white bread. Only people with more money could afford this bread. Very soon, white bread was considered better because that is what well-to-do people would buy.

Soon, manufacturers figured out how to use large rollers to grind wheat kernels and take out all of the germ and bran. They could sell this powdery white flour more cheaply than ever, and bakers started to use it to make fluffy cakes and pastries that everyone loved.

DID YOU KNOW…?

NUTRIENTS IN OUR FOOD

AT ABOUT THE TIME THAT WHITE FLOUR STARTED TO BECOME MORE POPULAR, something strange started to happen. People began to suffer more often with a disease known as pellagra. It caused their skin to get dry, red, flaky, and very sore. They had headaches and felt depressed and irritable. They had nausea and vomited a lot. If the disease got bad enough, they could die.

People in the southern United States, especially, suffered from this disease. It was thought that pellagra was caused by some kind of germ because it was more common in communities where people were very poor.

A very caring and observant doctor, **JOSEPH GOLDBERGER**, noticed that when these sick people were brought to hospitals, they stayed sick, but the staff who took care of them did not become ill. He had a theory that it wasn't germs that were causing pellagra. It had to be something else. But what?

DID YOU KNOW...?

NUTRIENTS IN OUR FOOD *(CONTINUED)*

DR. GOLDBERGER decided to try an experiment. He gave one group of sick people foods like vegetables, meat, and milk. The rest ate their usual diet, consisting mostly of cornmeal, molasses, and salt pork. The people who ate Dr. Goldberger's special diet got better, so he felt he had strong evidence that it was the food people were eating that was causing the illness. He had a very hard time convincing others to listen to what he had discovered. Unfortunately, Dr. Goldberger died of cancer before he could find the answer to the mystery.

It took a few more years, but scientists continued to study the "recipe" Dr. Goldberger had started to piece together. They eventually figured out that people were getting sick because they weren't getting enough of an important B vitamin, called niacin.

DR. GOLDBERGER

FRESH, HEALTHY FOODS
VS
PROCESSED, FOODS

GUESS WHERE NIACIN COMES FROM?
One important source is the parts of the grain that were being taken out to make white flour. People in the South were eating foods that were not rich in niacin, and that was the reason they got the disease much more often. Because there was no longer any niacin in flour, people's bodies started getting sick.

THE STORY OF
>>> ENRICHED WHITE FLOUR <<<

INSTEAD OF ADDING THE BRAN AND GERM BACK INTO THE FLOUR, food manufacturers decided to add some important nutrients by enriching the white flour. They mixed the flour with powdered vitamins and minerals, including iron and some important B vitamins. Eventually, the U.S. government made it a law that all white flour has to be enriched. This is the flour that most of us eat today. It is used to manufacture most of our baked goods.

Cookie scientists know that these kinds of changes happen to our food all the time. Food manufacturers make money when they invent foods that taste delicious, especially if they are so delicious that many people buy them. Sometimes, food manufacturers change the food in ways that taste great but are less nutritious for our bodies.

Many of the problems we have with being healthy today happen because of what we are not eating, not just because of what we are eating. A good cookie scientist knows this and learns about the important kinds of foods that need to be eaten in order to be healthy. This way, we can make healthy food choices most of the time and also make many delicious food choices.

THE STORY OF

>>> SALT <<<

BEFORE REFRIGERATION, SALT WAS USED TO PRESERVE FOOD, so it was critical for people to get salt. Without it, they could starve when food became scarce. Salt is so important for preserving food and preserving life that control of people's access to salt has been the cause of more than one war.

Requiring that people paid a tax on salt provided the Chinese government with enough money to build The Great Wall of China. Both the fall of the Roman Empire and the start of the French Revolution were related to rulers taxing salt too much. (A tax is money a government collects from its citizens to pay for the things the people in the country need.)

It took 2,300 years to build the Great Wall of China. It protected the country from invaders and kept trade routes secure.

Today, it stretches for about 13,171 miles along China's historic northern border.

THE STORIES
OF GREAT SCIENTISTS AND CHEFS

A MAN NAMED MAHATMA GANDHI lived in India and wondered if there was a way to free his country from heavy taxes on salt without a war. He decided he would "nonviolently" break one of the British Government's laws he thought was unfair. He went on a 240-mile march from his home to the Arabian Sea. There, he picked up a piece of salt out of the mud at the edge of the sea. That is how he broke the law!

Salt was plentiful along the coasts of India, but the British leaders had made it unlawful for native people to gather it or sell it. Instead, salt was shipped from England and then sold at high prices that the British government controlled.

Because salt was so important to keep people healthy in India's warm climate, Gandhi thought this law was especially unfair to people who were poor. The Salt March was just one example of the way Gandhi thought like a chef. Throughout his life, he continued to help people think of non-violent ways to protest laws they disagreed with and wanted to change.

Whoever controls the salt supply can control an entire country. Just remember that the next time you forget the salt in your cookie dough!

THE STORY OF
>>> BAKING SODA <<<

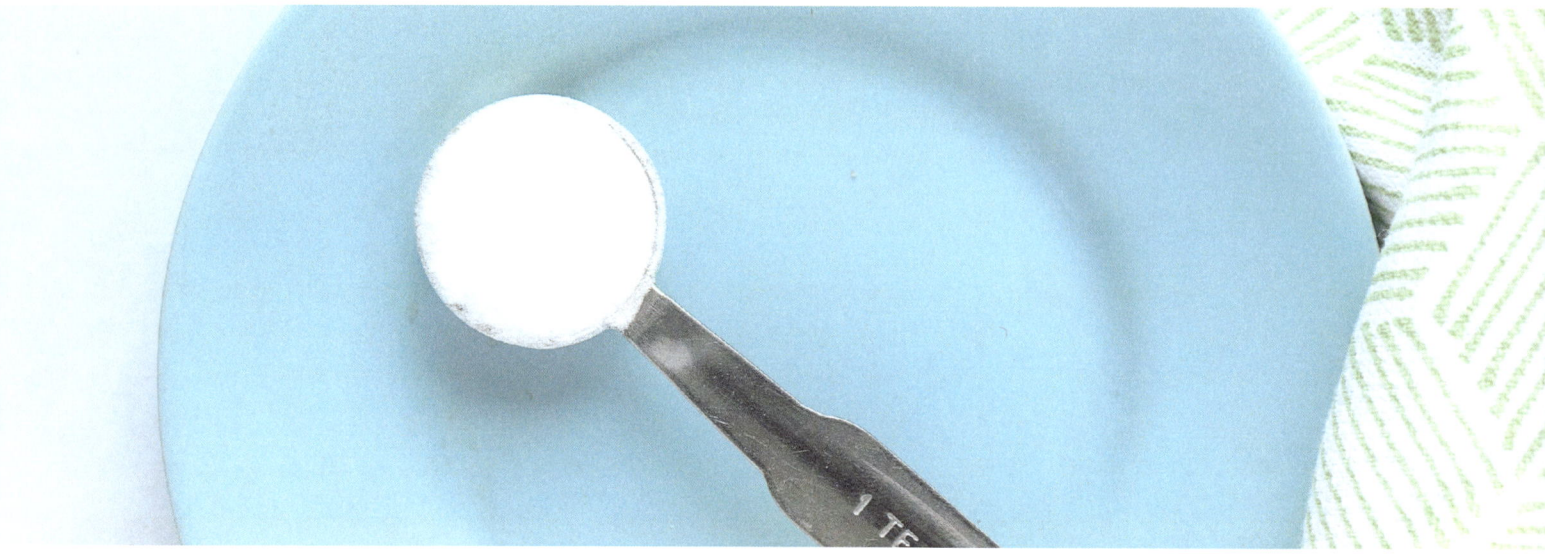

TWO CHEMISTS, NICOLAS LEBLANC IN 1791 AND VALENTIN ROSE IN 1801, independently developed processes to create sodium bicarbonate. However, it wasn't until 50 years later that bakers began using it widely. Before then, making a cake rise meant adding yeast to the batter and waiting a couple of days for it to rise—hoping it wouldn't fall flat before baking. And since white flour wasn't commonly used yet, let's just say their birthday cakes were nothing like the ones we enjoy today!

Once baking soda was available, people invented all sorts of uses for it. They discovered you can use it to kill cockroaches, polish silver, extinguish fires, and eliminate smells from old musty books. Today, sodium bicarbonate is used to clean teeth, save patients who are in cardiac arrest, remove paint and rust, ease heartburn and indigestion, and as a household cleaner.

BAKING SODA IS TRULY ONE OF THE MOST USEFUL CHEMICALS IN YOUR HOME.

THE STORY OF CHOCOLATE

ANCIENT MAYANS are believed to be the first people to grow and harvest cacao, which is the plant chocolate comes from. Cacao beans grow inside large pods on cacao trees. The Mayans ground up the beans to make a powder, then added water to make a chocolate drink. Unlike the hot chocolate we drink today, it did not contain any sugar.

CHOCOLATE WAS VERY VALUABLE—EVEN MORE VALUABLE THAN GOLD!

Therefore, only members of royalty and rich merchants drank it. Sometimes, it was used as a reward for soldiers returned from battle. Cacao (kakaw in the Mayan language) was an important part of the Mayan and later Mesoamerican cultures. The beans and seeds were used as money. Because the cacao tree grows best in a tropical climate and in the shade of larger trees, most of the cacao beans grown worldwide today are grown on very small family farms of just a few acres. Once cacao is fermented and dried, we call it cocoa.

THE STORIES
OF GREAT SCIENTISTS AND CHEFS

FOR HUNDREDS OF YEARS, cocoa made from roasted cacao beans was only consumed as a drink, but in 1850, **JOSEPH FRY** discovered that you could add cocoa butter (a fat that comes from cacao beans) to cocoa powder and turn it into a solid that could be shaped. This introduced a lot of fun chemistry experiments as different chefs found unique ways to use chocolate. They could mold it, flavor it, and dip other things into it.

In 1857, a pharmacist named **JEAN NEUHAUS** started to dip medicines for his customers into chocolate to make the medicine taste better. Amazingly, it wasn't until 55 years later, in 1912, that his grandson, **JEAN NEUHAUS II**, had an idea to start dipping centers of sweetened candy into the chocolate instead of medicines. He called this a "praline," and it set off a storm of other candy and candy bar inventions. It's hard to believe that we have only known how to make chocolate into candy for about the last 100 years!

COOKIE SCIENCE

AND THAT BRINGS US BACK TO THE BEGINNING, when Ruth Wakefield started cutting up pieces of a flat chocolate bar to add to her cookies and published her first Toll House Cookie recipe in 1938.

If you are a cookie baker, you are also a chemist, a mathematician, a creator, and a scientist. Baking teaches you about math, measuring, and blending flavors. It even teaches you what happens when you make mistakes. All chefs and scientists make mistakes. It's an important part of the process of learning, creating, and discovering new things.

NOW THAT YOU ARE A COOKIE SCIENTIST, you can experiment just like Ruth Wakefield. You can test new recipes in the kitchen and write them down like Auguste Escoffier. You can think about ways we have been getting things wrong and consider whether there's a better way like Joseph Goldberger and Mahatma Gandhi. You might use your knowledge to create art that inspires other people the way Caroline Shaw Brooks did. Most importantly, you can share what you have learned with other people like Edmond Albius and the men who developed the first insulin therapy did. Others will be able to try your experiments for themselves and make new improvements.

What will you put into your next batch of cookie dough that the recipe doesn't call for? What will you leave out? Someday, you might invent something completely new that no one has ever thought of before. Doesn't that make you feel like one smart cookie?

The Stories and the Science | 51

THE AUTHOR

>>> LYNNAE ALLRED <<<

LYNNAE ALLRED is a cookie-baking grandma who loves to examine spider webs and seashells, jump in puddles, and go on fun vacations with her grandchildren. Scan the QR code below to visit her website, LynnaeAllred.com/cookies, where you can download more fun cookie-baking projects to try on your own or with your family.

www.ingramcontent.com/pod-product-compliance
Lightning Source LLC
Chambersburg PA
CBRC091202070526
44583CB00008B/176